ABOUT THE AUTHOR

Master in Information Technology Governance, graduated in Information Systems with specialization in Systems Management and Information Technology from the University of California - Berkeley; MBA in Project Management from Fundação Getúlio Vargas - FGV. He has strong experience in budget control in the areas of projects, IT and quality, business profitability, cost analysis, economic and financial feasibility studies. Experience in administering service contracts. Experience in Business Management, getting involved with the administrative, financial, commercial, production and logistics areas, participating in investment policies and coordination of corporate plans, strategic analysis and definition of market positioning. Strong performance in the Commercial area, encompassing feasibility and prospecting for new business, market analysis, sales performance, customer satisfaction, marketing plans, structuring of strategies and general policies at the corporate level, definition of guidelines and strategic actions. Solid global knowledge of business, having relationships with partners in several countries, distributed in the commercial, corporate / executive segments. Career developed in the Information Technology and Projects area. Management of multidisciplinary teams, monitoring progress and reporting results. In IT Governance, ensuring the interface with the business areas. Strong career in strategic function with internal and external teams. Physical and financial control of projects through tools and performance indicators, feasibility analysis, cost reduction.

Adherence review and process improvements. Project management for national and international clients.

INTRODUCTION

The content of this book is based on the experience lived along a professional trajectory marked by different opportunities, varied cultures, different verticals, among other challenges that enabled a path full of variety and unique learning. The scenarios experienced along this path allowed the formatting of materials marked by real examples and not just written theories, and which in a simple and light way allows transferring this knowledge to any level of professionals..

The PMO (Project Management Office) or Project Office, is the unit that centralizes and coordinates project management, through guidance and support, according to the organization's maturity levels.

The first objective of an organizational system aimed at the execution of the company's projects - Projected Structure, is to identify the level of maturity through the classification (maturity survey), in addition to knowing the advantages and disadvantages of adopting this type of structure.

This booklet works as a "mental map" that starts with the adoption of the first project to be managed and, from the "top" of this need; identify the best model according to the company's maturity in relation to the processes existing until then (it would be the intention of the Project Office and its feasibility according to the reality of the results obtained by the known processes).

The objective here is to recognize whether the company feels the need to formalize the way it manages its projects and the respective exchange of information between projects; and how much are you willing to change to implement continuous improvement in project management.

METHODOLOGY

The mind map constructed in this book is intended to be observed individually, reading only the images (a dream from the beginning of reading practice, isn't it? Looking only at the images). That's right: the primer speaks for itself. However, to support the logical sequence, always based on the best practices suggested by the PMI (Project Management Institute) and presented in the PMBoK; for each new step of the flow there is an explanatory text.

The best scenario would be to separate the texts (part I) in their logical sequence and all the figures (part II) with their respective sequences, so that there are two reading formats that can be understood in their explanation (part I) and then illustrated by its figures (part II).

Now that you understand the parts separately, let me just say that this is not how I did it! Each figure has its macro-explanatory part, which supports the common understanding even if the READER has no experience in projects..

Project Management Processes – Pratical Conduct Guide

SUMMARY

1. **RESEARCH AND MATURITY CLASSIFICATION**
   1.1. I NEED A PMO?
2. **DO I HAVE A PROJECT TO MANAGE?**
   2.1. THE 5 PHASES OF THE PROJECT
   2.2. WHAT TO DO NOW?
   2.3. WHAT IS PROJECT?
3. **1ª PHASE: STARTING THE PROJECT**
   3.1. HOW SHOULD I PLAN?
   3.2. PROJECT PERCEPTION
   3.3. THE INTERESTED PARTIES
4. **2ª PHASE: PLANNING THE PROJECT**
5. **3ª PHASE: EXECUTING THE PROJECT**
6. **4ª PHASE: PROJECT MONITORING AND CONTROL**
7. **5ª PHASE: PROJECT CLOSING**
8. **LESSONS LEARNED**
9. **A BRIEF OVERVIEW OF COSTS**
10. **WHAT IS THE BEST WAY TO MANAGE PROJECTS?**
11. **WHICH TEAM IS THE BEST?**
12. **AM I A GOOD LEADER?**
13. **FINAL CONSIDERATIONS**

1. RESEARCH AND MATURITY CLASSIFICATION

The research and maturity process presented below tries to represent the model of questions that will direct the best effort according to the company's reality. The term "Effort" is the degree of control in the processes to be carried out when we adopt a methodology to be followed within the company, so that the information, historical data, documentary models, among other tools that will be used to carry out a given project.

Colors and formats were adopted that the reader will easily assimilate to follow the flows suggested in this booklet.

Some formats end with titles that are part of the second volume of this booklet; but this does not prevent or leave incomplete the objective proposed here (ex: MATURITY LEVEL 1 is an "output" presented in a flow). We will not explain maturity levels in this material. This is a deepening that is part of the micro content that we propose as study material: We start with a macro flow and each sub flow will be visited after understanding the previous level. Believe me, this top-down model allows you to understand this broad and all-too-common problem, even if it's not labeled in people's everyday lives;
in a simple and clear way. GOOD READING!

1.1. I NEED A PMO?

The PMO (Project Management Officer) is the sector responsible for the centralized management of the project portfolio. But what is the "project portfolio"? When we work for projects. It is an organizational system aimed at carrying out projects. Thus, if each sector has a certain budget, a master plan defined and aligned with the organization's strategic objectives; we can consider each desired investment as an idealized project to be carried out. These initiatives, these desires, these intentions of each area, will be part of the project portfolio and will have their classification and prioritization listed based on the company's strategy. To compile information about the idealized projects is to try to facilitate the decision-making of the executive committee in the approval of this initiative. The initiatives are analyzed and qualified. This must be done in a structured way and with templates defined for this purpose; taking as an example GUT Matrix, 4x4 Matrix, RICE Matrix among others. The PMO is responsible for standardizing project management, promoting transparent communication, monitoring and reporting the indicators of ongoing projects, taking care of resource allocation, monitoring progress and, above all, keeping the lessons learned knowledge base updated..

To have or not to have a PMO in the company is a decision that can be made by evaluating the company's scenario. The degree of maturity in relation to the treatment of projects that the company has. And the suggestion here is to apply a "framework" that helps to identify whether the company would benefit from this structure or not. Remember that this change must be very well communicated as it impacts the entire organization. Therefore, it is essential to identify the current scenario regarding the degree of maturity of the company in the issue of project management.

It is also important to remember that the proposal described here is not a general rule to be adopted by all organizations; but rather a model that serves as "north" for any company to identify in which context it is inserted for the adoption (or not) of the famous project office, or PMO. But be careful! It is essential to understand that having a PMO is having a budget available for this purpose and, correspondingly, achieving the expected positive result with this structure. Managing projects is managing schedule and costs; aiming at reducing project costs, reducing project failures and meeting customer expectations (internal).

It is not an immediate result, as it is about the dissemination of best practices. The results are formed along with the cultural change of that system..

Project Management Processes – Pratical Conduct Guide

Project Management Processes – Pratical Conduct Guide

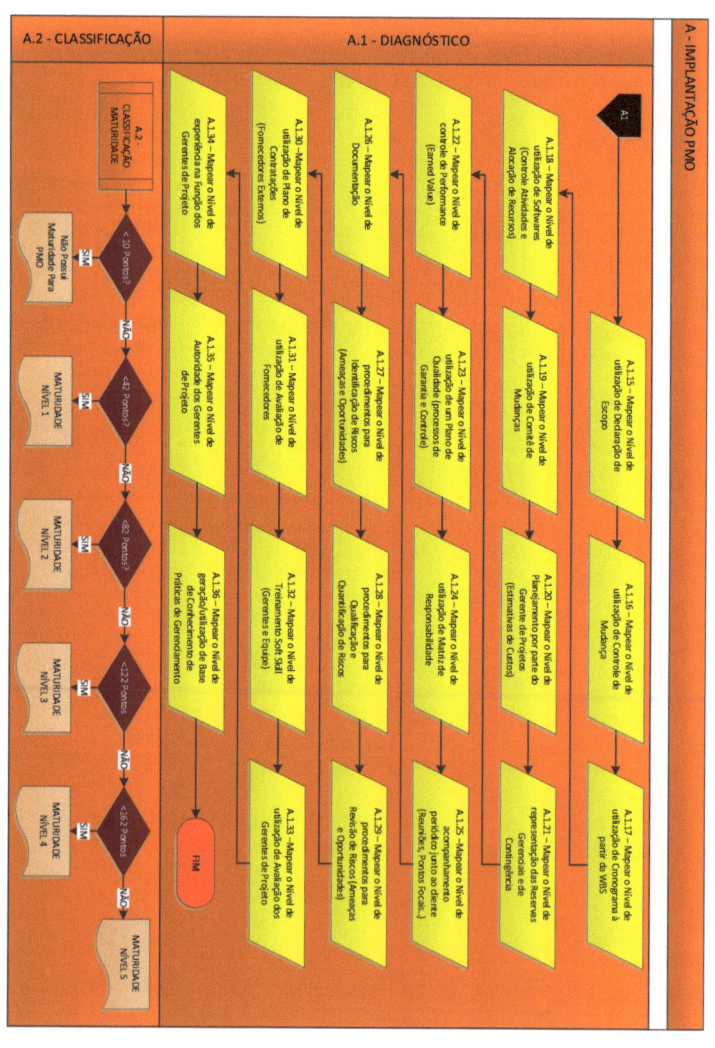

Project Management Processes – Pratical Conduct Guide

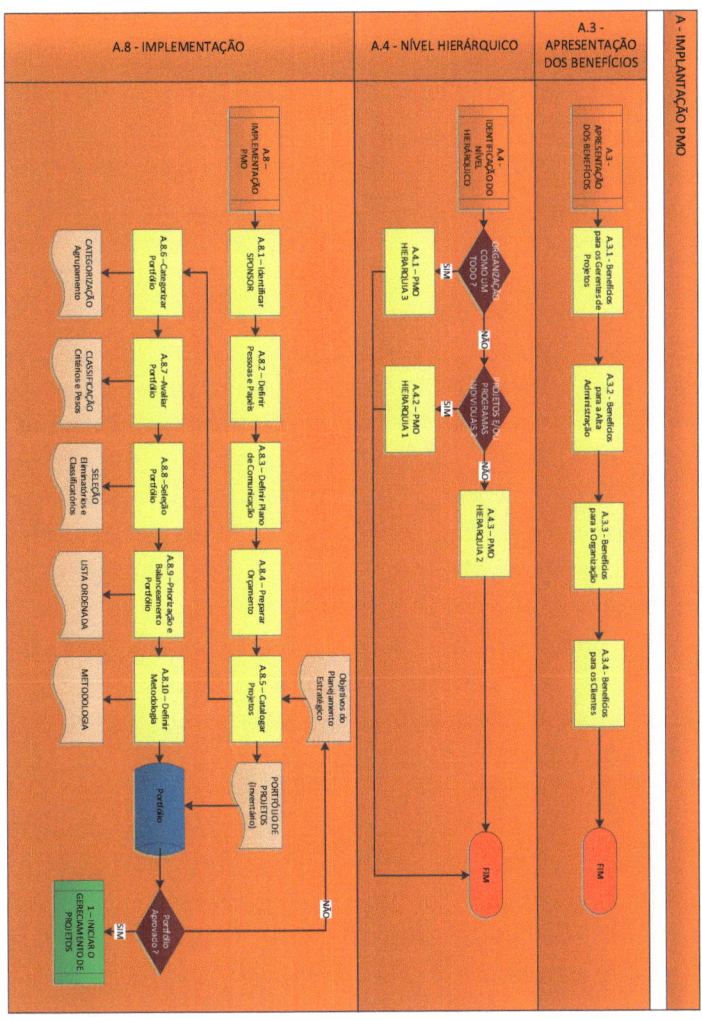

2. I HAVE A PROJECT TO MANAGE?

By definition, a project is a temporary effort undertaken to create a unique product, service, or result; with defined beginning and end. They are planned, executed and controlled by people, limited by well-defined resources (labor, equipment, services...), progressively elaborated.

According to PMBoK, project management is the application of knowledge, skills, tools and techniques, and translates into best practices distributed in 5 groups of processes (Initiation, Planning, Execution, Monitoring / Control and Closing); which function as a live PDCA (Plan - Execute - Check - Action) throughout the entire project in progress. And as a PDCA, the distribution of processes suggested by the PMBoK to be monitored and revisited are considered the best project management practices, providing a mechanism for reflection and reassessment by activity packages, mitigating risks, anticipating and eliminating problems, forcing a look broader to achieve the ultimate goal.

Based on these suggestions, each company will be able to adopt its own methodology for implementing projects, according to flows and documents it deems relevant for the application of what would be the best management practice..

There are currently 47 processes distributed in the 5 process groups. Depending on the type and / or size of the project, it may happen that we do not use all the processes described in PMBoK. So never say "PMBoK Methodology"! These are "best practices". The methodology you can contribute to create in your organization.

The lived experience leads us to understand the known definitions. As a "project", even considering the same business vertical, each project is unique, it is exclusive. There is no "copy and paste": It is very interesting that if we stop and think hard, we will surpass 50% of an ongoing project and we will still be identifying opportunities that could have been considered, revisiting lessons learned, identifying new requests for changes, planning an improvement of product; requirements not initially identified and which are expected. And for sure complaining about every moment we didn't get it right the first time; but admired for the surprise and complications that a project presents along the way..

2.1. THE 5 PHASES OF THE PROJECT

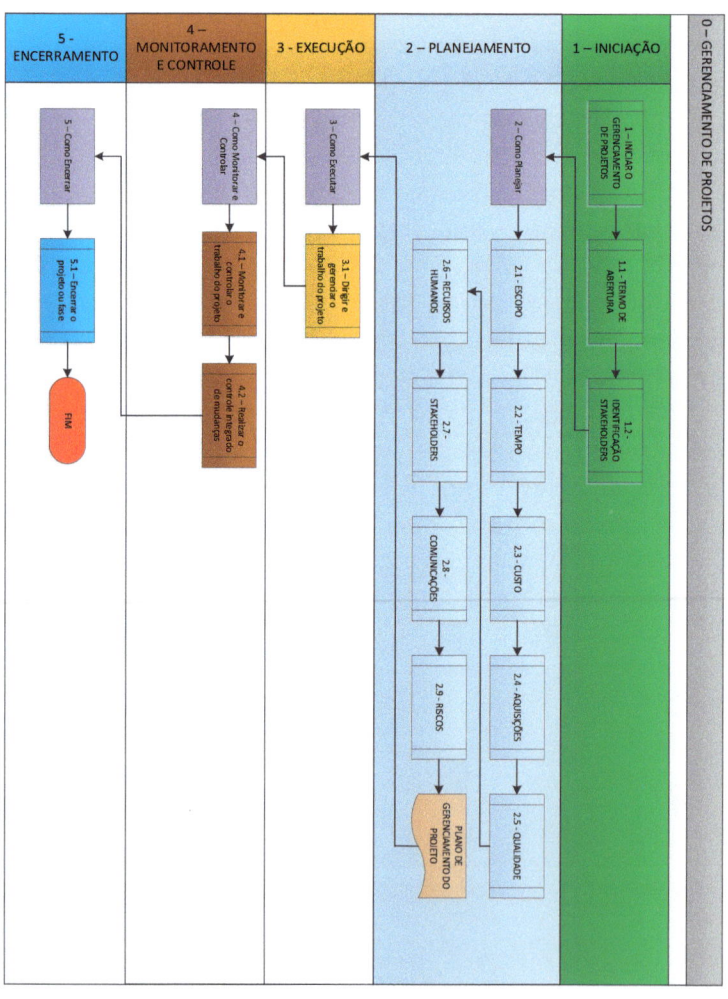

2.2. WHAT TO DO NOW?

The future does not depend on chance, divine providence or circumstance (DeMasi, The Creative Leisure). We are responsible for designing our own future.

The vision presented in this booklet presents a structured sequence, facilitating the thinking, the mental reasoning that all of us, in a way, do when we are faced with a problem. Often we cannot abstract correctly, but; it is inevitable that we must go through all project management processes implicitly..

In the flow presented in this booklet, we suggest some "outputs" (reports / information) as well as the moments when processes go through a database. All of this, if well used, guarantees the history of the "Lessons Learned" that can be obtained at any time, in a structured way, to serve as support and contingency in another project.

2.3. WHAT IS PROJECT?

The project exists because of a need. It must have a name, a title that represents its reason and benefit. Certainly, in a company, each and every project is aligned with the company's Strategic Planning. We must ask ourselves:

- Why should the project be carried out?
- Which or which ideas should be addressed?
- What limitations must be respected?
- Is there connectivity to another project?
- What is the scope (end product)?
- Limitations: What must be delivered (within scope or positive scope)?
- Limitations: What should not be delivered (out of scope or negative scope)?
- Descriptions and documentation that should be considered as project delivery?

It is the analysis of the current situation and what is intended to change, causes and effects, as well as possible resistance to the project.

All assumptions and RESTRICTIONS must be analyzed and their impact on the project scope.

Project Management Processes – Pratical Conduct Guide

The scope is the larger set that describes the final product and the entire structure of activities necessary to achieve this goal.

3. 1ª PHASE: STARTING THE PROJECT
(The organization is committed to an objective)

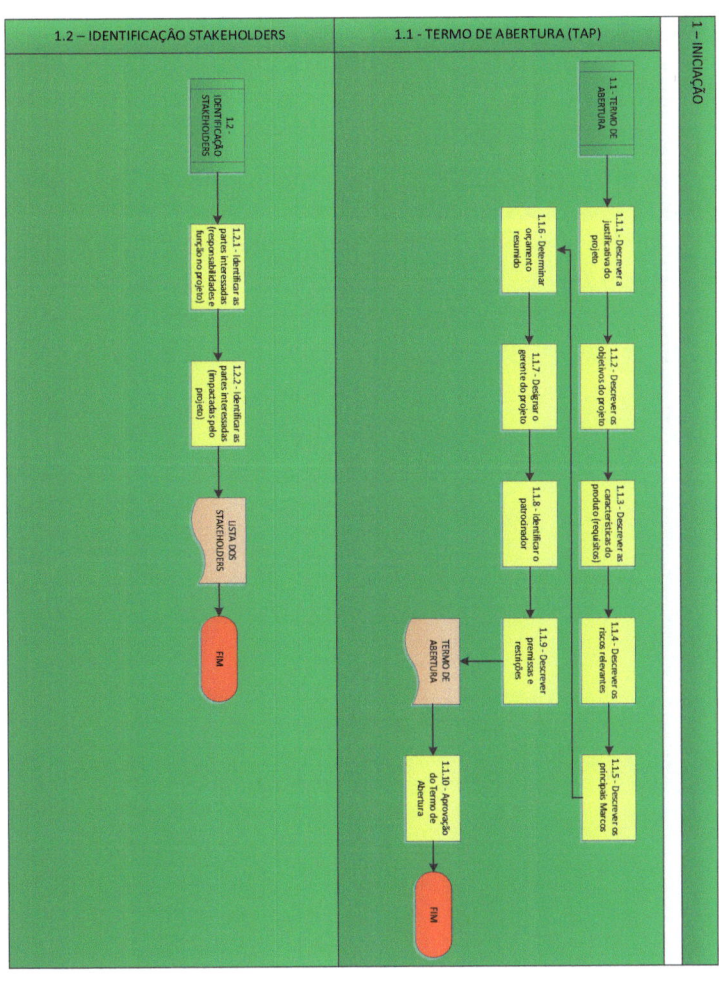

3.1. HOW SHOULD I PLAN?

In particular, the planning phase is actually the final project and execution is just the means! Practice reveals that execution becomes a succession of new plans derived from the monitoring and control phase.

Many project managers believe that they are applying risk management, problem identification and application of different techniques to achieve the lowest cost problem solution to meet the end customer's expectations (whether it is a person or a sector of the company itself - internal; or another company - external). But the truth is that they are moving towards the best possible contour within a project that is already underway..

An ongoing project receives internal and external influences that are often not mapped out in the Inception and Planning phases. For the reasons that required the project, these first two phases are done quickly, as you want to start the project and bring the answers. This is a common mistake!

Communication is a common problem in people's lives. We always listen and prioritize our perception, when in fact, we must place ourselves in the perception of "the other". We are so concerned with understanding what the customer needs that we listen to ourselves more than the customer. And as phases are delivered, problems arise and with them the best alternative solutions to the problem at hand: This is now a fact! The problem does not require perception as it has already occurred!

3.2. PROJECT PERCEPTION

The figure below symbolizes the number of actors involved in a simple project, according to their perception, bringing a classic illustration of the different forms of interpretation:

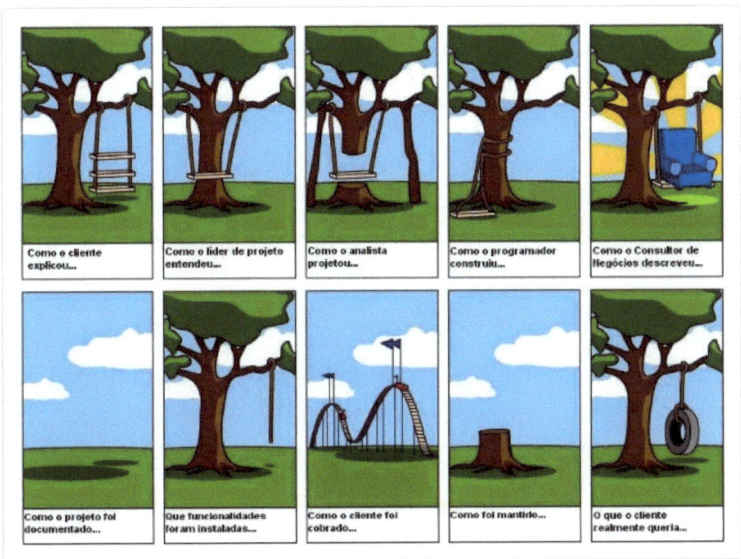

Source: Web

The more we dedicate ourselves to gathering all the requirements, we identify the parties involved (directly or indirectly, but that will somehow be impacted by the project) and plan the project, we will be more assertive in our deliveries and expected satisfactions.

3.3. THE INTERESTED PARTIES

In my experience, I reinforce the care with the "stakeholders" (or Interested Parties). Stakeholder management is really the area of knowledge that requires dedicated care. Stakeholders influence and are influenced by ongoing issues and need to be involved in varying degrees of frequency and intensities throughout the project lifecycle. Ignoring this caution is a big mistake on the part of the Project Manager.

Every project has a beginning full of expectations and the relationship is one of total partnership. But as the limits of the comfort zones are crossed, the interferences of the ongoing project can cause discomfort and even be seen as a threat: Several factors are related to this discomfort, such as process improvement and optimization, information integrity, traceability, feasibility of information for almost instantaneous strategic decision-making, among others.

Mapping and identifying stakeholders is essential to mitigating risks, allowing you to work with them from the beginning to explain the change that is to come. These different parties involved have common interests as a starting point. In projects that involve many stakeholders, the overlapping of individual interests over collective ones can lead to attrition, weakening and destruction when resources are scarce or poorly managed..

In medium and long-term projects, the fierce competition between the parties in defense of their private interests promotes the inability to fulfill commitments, limits their progress and even prevents their completion in the absence of the leaders..

Management of Stakeholders is, in a way, managing the sustainability of the project, as it aims to legitimize its activity, respect the limits and restrictions of the interested parties and the systems in which they participate and, in short, the feasibility of projects to avoid the catastrophe of disorder.

Thus, all interested parties must be identified. And, make no mistake; when we talk about interested parties, we are not limited to the company's employees. These are already a challenge and need to be classified in the way they will participate in the project. We cannot forget that, within the organization, the interested parties are not just those who operate in the information systems (or users). But also those responsible for tasks that will have their results measured in these information systems.

For example, when many deployers are talking about inventory, they forget to include those responsible for manual tasks within the inventory sector in the project.

This workforce that lives the reality of storage and movement of products and cargo knows the dynamics more precisely about, for example, WMS (Warehouse Management System). Hearing only the leaders before an implementation project is a common mistake, believe me, and it shouldn't be a practice. A large part of the manual workforce has a lot to contribute in defining processes.

It is important to remember that each stakeholder has a unique expectation of the project. We must learn to deal with these differences, because during a project, what matters least is our understanding of the scope; but rather the expectation of the "other". And understanding the expectations of the "other" is the great challenge of management.

The worksheet below demonstrates a Stakeholder Matrix that is critical to identifying each human resource within the context of the challenge ahead. In addition to the list of Names and Emails to identify a particular resource, the Position identifies its "power" in the organization.

We use 3 (three) categories:

1. **Supporter:** One who accepts and encourages the change to come.
2. **Neutral:** Neither supports nor opposes. But he may change his mind if the change presents any difficulties.
3. **Resistant:** Does not accept the project.

Power and Interest are also factors that should be considered and should be classified as High and Low. High power resources when the Supporters are part of the team that wants the final result of the project and will help in the resolution of conflicts. But it's no use if they have little interest in the project. At the other end are the resistant.

Note that the extremes are on one side the "Supporters" with High Power and High Interest and on the other side the "Resistants" with High Power and Low Interest! Both need to be managed in a particular way. Describing the strengths (positives) and weaknesses (negatives) of each stakeholder, carefully mapping each individual to know exactly how they should be handled throughout the project lifecycle is a valuable activity..

Don't miss the opportunity to leverage this information for the benefit of the success you want to achieve. Projects are carried out by people. There is a triple "PPP" that supports the business: Product (which refers to the software(s) used, tools, etc.), Process (which refers to the procedures used in the execution of the activities aimed at the business results) and finally the People! This last "P" is decisive for the success of a project. That last "P" is what guarantees success or failure. Knowing how to deal with this last "P" is fundamental and requires skills.

These skills can be acquired throughout the acquired experiences, through courses and complementary reading.

This table will indicate 3 stakeholder profiles that should have their efforts managed among Supporters, Neutrals and Resistances:

- Manage closely;
- Keep yourself satisfied;
- Keep informed.

If the reader hasn't noticed, a lot of what we are talking about here are 2 (two) very important management areas (in addition to the Stakeholder Management area): We are talking about Communication Management and Risk Management!

A project is actually the care of these 10 (ten) areas that are the project's knowledge areas.

Project Management Processes – Pratical Conduct Guide

STAKEHOLDER	EMAIL	POSIÇÃO	CAT	PONTOS (+)	PONTOS (-)	PODER	INTERESSE	ATITUDE
			Apoiador			ALTO	ALTO	Gerenciar de Perto
			Apoiador			ALTO	BAIXO	Manter Satisfeito
			Apoiador			BAIXO	ALTO	Manter Informado
			Apoiador			BAIXO	BAIXO	Monitorar
			Neutro			ALTO	ALTO	Gerenciar de Perto
			Neutro			ALTO	BAIXO	Manter Satisfeito
			Neutro			BAIXO	ALTO	Manter Informado
			Neutro			BAIXO	BAIXO	Monitorar
			Resistente			ALTO	ALTO	Gerenciar de Perto
			Resistente			ALTO	BAIXO	Manter Satisfeito
			Resistente			BAIXO	ALTO	Manter Informado
			Resistente			BAIXO	BAIXO	Monitorar

But don't think that the work with stakeholders is over! Stakeholders are not just limited to the organization's internal resources. When we talk about Stakeholders, we are talking about shareholders, senior management, board of directors, employees, suppliers, customers, banks and other creditors, regulatory institutions, investors, public bodies, the community in general... These must also be analyzed.

There are many examples of when these stakeholders are not considered, the project can have a high cost due to scope change, missed deadlines, new hires or contracts, among other affected areas of knowledge; or even be driven to failure.

We have several examples of projects that had their course changed during their execution. One of the most famous cases was the Belo Monte dam in Pará (Brazil). Since its inception, the Belo Monte project has encountered strong opposition from Brazilian and international environmentalists, some local indigenous communities and members of the Catholic Church. This opposition led to successive reductions in the scope of the project..

Good planning is the key to success. Think about looking at each activity at the lowest possible level, one that is worth the effort of being monitored: it's like making a point and looking around that little dot! Everything related to materializing that little point. Its perspective, size, color, texture, etc., and all the factors around this point that can interfere, favor, prevent, depending on the point being delivered to the customer at the desired time.

If you've never understood the expression "getting out of the box", now is the time. Looking "from above" and expanding the vision around the project is a valuable exercise. You begin to notice obstacles and opportunities that must be considered. This limit is determined by you! And hence comes the issue of maturity described in Chapter 1.

The size and time devoted to projects are in accordance with the organization's capacity to deal with this type of structure. The most important thing is to know that planning is the most opportune moment to make a successful project feasible. It's time to explore possibilities, understand nuances, plan, plan and plan. Good luck and sucess!!

4. 2ª PHASE: PLANNING THE PROJECT
(Defining the rules of the game)

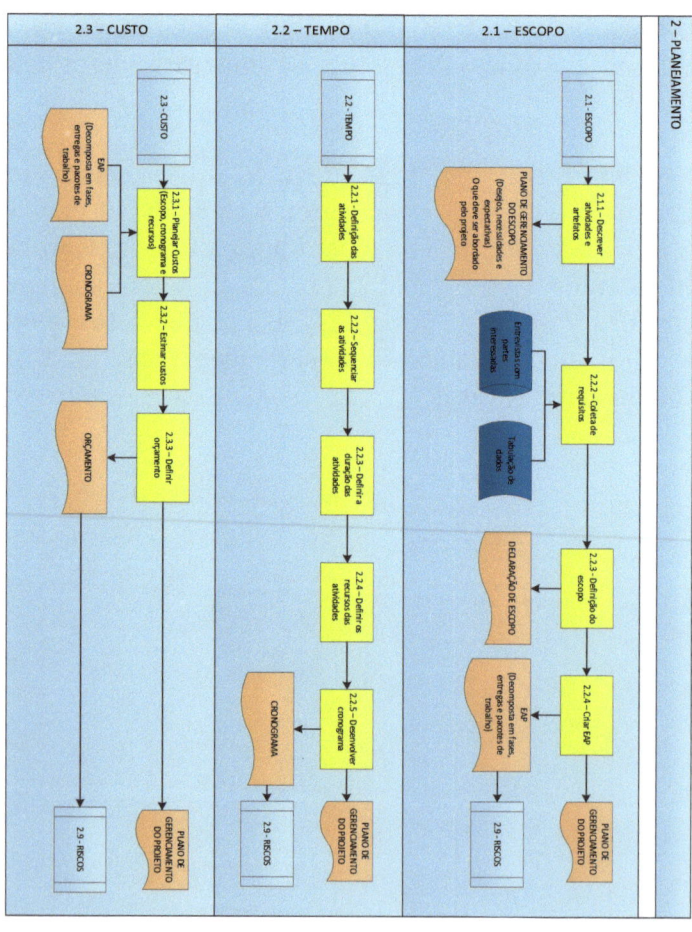

Project Management Processes – Pratical Conduct Guide

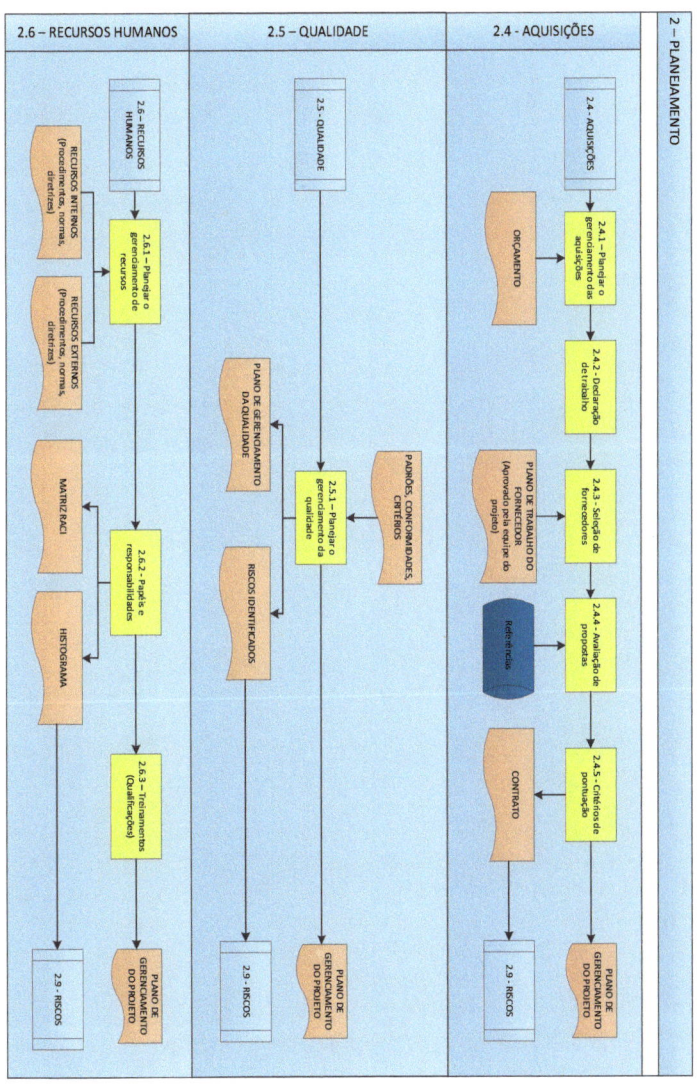

Project Management Processes – Pratical Conduct Guide

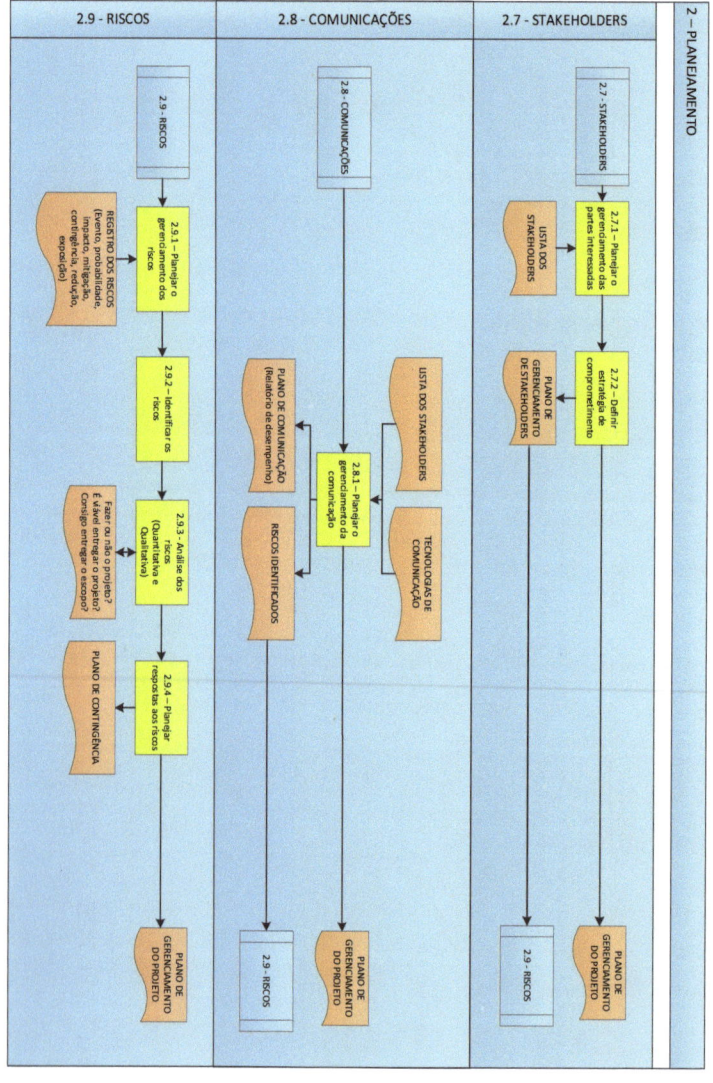

No planning will be irreplaceable! Yea! It's an affirmation! In the same way that a project takes temporary effort, executed by people and produces a unique result; a given plan is inherent in a single project. Much can be enjoyed! But believe me: re-plan the next project! And then you understand the value of "Lessons Learned".

Remember the design of the tree with the swing? What was the customer's expectation really? And the little point? The sum and totality of the little points is exactly the scope of the project. All small points must converge to the equality of the final point which is the Scope of the project.

Understand the scope, the final, unique and exclusive product... And look around! Imagine if I say I want a glass of orange juice. What is the cup size? With sugar or without? Ice? Ordinary orange ?

But I also don't need to think about presenting the cup on a silver tray, with napkins, fruit ornaments on the edge, etc... This is another type of problem in projects that we'll talk about in another sequence in this literature (very important by the way: the limits from the project).

The factor here in this provocation is to understand that we need to know what you want and create the paths to build this product... And it can literally mean building paths, as one of the little points of the project: My project is to build a rocket and deliver in the center of the city of

São Paulo. Let's think? Part of this project is to think about the route to take the rocket to the center of São Paulo. If I stay focused on building the rocket, being with the best NASA team, considering material, suppliers, safety, aerodynamics, fuel, weight, etc and etc and forgetting that there is "downtown São Paulo" for the rocket to end there; I'm going to have a serious problem; because I haven't evaluated how to do this yet and this lack can mean a lot.

This could mean huge costs and even a tragic end. It may be better to build the rocket in downtown São Paulo than to transport it there, it's not true?

This is planning! That's the real challenge! The greater planning effort implies less effort in execution, monitoring and control; as well as lower incidence of changes throughout the project.

5. 3ª PHASE: EXECUTING THE PROJECT (Executing the created plan)

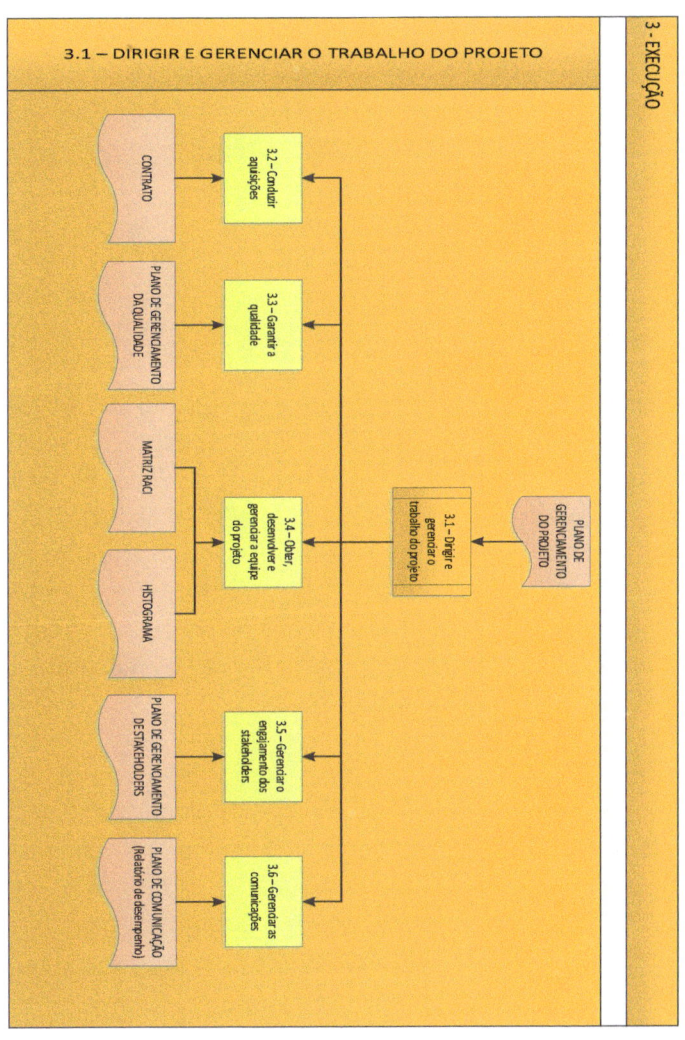

At this stage, the beginning is beautiful and wonderful, as we have the plan in hand, resources are defined and all areas of knowledge are consolidated through their respective plans (Integration, Communication, Stakeholders, Scope, Time, Quality, Acquisitions, HR , Costs and Risks). Just follow along... But it's never like that! And this is what most excites project managers driven by their passion for the profession.

Here, expectations for the new are vibrant! Will happen! It's already happening! Several characters are part of the change to come.!

Project execution is carrying out the work defined in the Project Plan, updating and controlling. And that involves absolutely everything around this idealized point: People, equipment, weather, day of the week, week of month, month of year, year of life... and the flapping of the butterfly's wings on the other side of the continent! Everything changes! Managing projects is managing people. And people are other small points that impact and are impacted by the project.

And the changes arrive that must be measured and evaluated. To understand a change, at some point we look at the small point and identify a deviation in any attribute of that point, inside out or outside in. In fact, during project execution, we never take the next step without properly observing where is the foot. This is the monitoring and control of the project. The closer we are to these scenarios, the faster and with less effort we can get around. And speaking of running, checking, bypassing and planning again; we are carrying out our PDCA.

Execution is nothing more than putting into practice the initially defined plan. The better the plan, the smoother the execution. I usually say that during execution, each step taken should be reread calmly; look at the new state, but also look at the previous state. It's like walking on unfamiliar terrain with shaky ground. Step on! Look carefully. Understand carefully each change.

6. 4ª PHASE: PROJECT MONITORING AND CONTROL
(Monitoring the plan created)

Following the project is revisiting the plan, derived from the scope: that is, always revisiting the scope and verifying that the plan is really adhering to it. All plans for all knowledge areas are reviewed and changes identified and recorded.

These changes need to be approved and the results of them become part of the Project Management Plan. Change Requests must be formalized, with their cause described and respective responsibilities assigned.

What is the cost of this change?
What is the impact of this change on other stages of the project?
What is the impact of this change on project stakeholders?

Of course, the project plan is not static!

Managing Projects is to carry out a continuous PDCA throughout the project cycle.

Project Management Processes – Pratical Conduct Guide

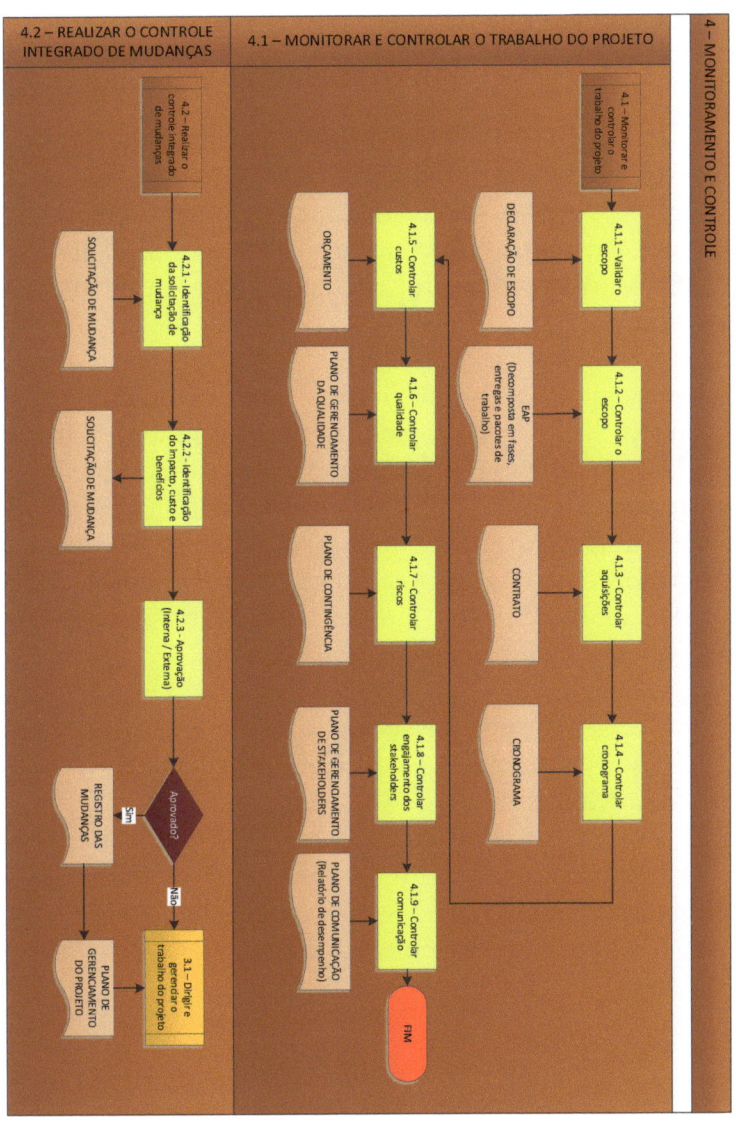

7. 5ª PHASE: PROJECT CLOSING
(Closing of the created plan)

Each phase of the project, planned deliveries, milestones, products, modules, etc ..., must have its "closure" according to the methodology agreed between the interested parties. We can also call it Homologation. The sum of all approvals means the approval of the project as a whole.

As mentioned at the beginning, the methodology is relevant to the model adopted by the institution; that can range from simple to more complex. We return to the issue of maturity. And not just the company. But the resources that are assigned in project management activities. This is another approach that makes sense in the literature when we talk about the "fundamental skills of the project manager" and their respective competencies.

The project's Closing phase aims to complete the work and, similarly, to complete each plan generated by each area of knowledge.

The project was approved by a sponsor. This is the same representative who must acknowledge the closure, the consent that your expectations have been met.

Project Management Processes – Pratical Conduct Guide

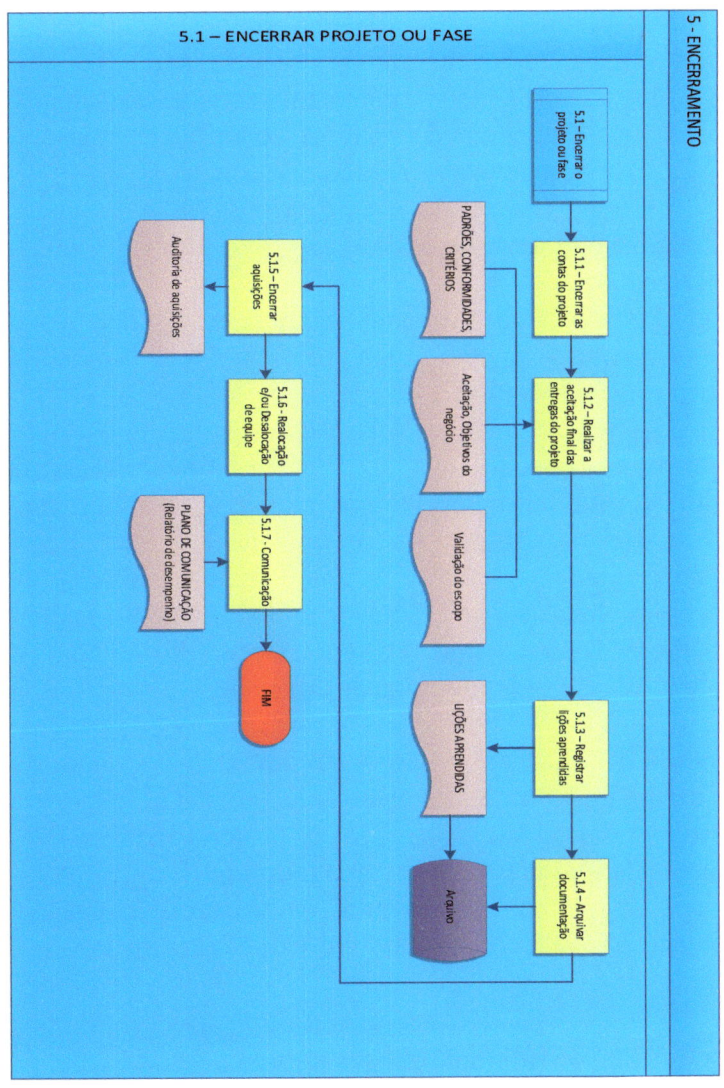

8. LESSONS LEARNED

However, throughout the project lifecycle, some fixes were made due to unforeseen situations, not noticed when the project was conceived. All of them had reason(s), impact(s) and action(s) to resume the course.

We realize here that we have a story that can and should be described in such a way as not to fall into oblivion and have this event re-presented in another project. So the question is: how so? Isn't each project unique? Why should I keep this information?

Each project is temporary and unique. It is true! But this idealized product or service that gave rise to the project is surrounded by a reality that is no longer unique: People, physical place, related technologies, administrative processes, legal factors, society, among others. And it is for these reasons that changes occur throughout the project. It's the old "moving car tire change".

Recording "Lessons Learned" is a very important historical basis to avoid the same "mistakes" that were corrected during the course of the project. How to record lessons learned is part of the methodology adopted and has several suggested formats and, once again, it meets the maturity of the company and the value it represents in the effort.

Managing projects is a process of continuous learning, a constant PDCA in the way of managing, a reinvention of posture and behavior in the face of the different scenarios that are presented to us and at each step taken in the execution of the initially prepared plan. The record of lessons learned should not be limited to situational points directly linked to the execution of activities in progress; but also the situations experienced throughout the career, motivated by differences in organizational models and cultures, profiles, mission, vision, values and audiences involved in that specific time and place.

Managing a project is like driving a car you already know and believe can ensure it runs smoothly. However, the conditions of the road you are on, the geography of the terrain, the climate and weather conditions, the available signaling, the available support, among others; go beyond the limits of what you could guarantee and start driving what you know in an unknown setting: The good driver drives under constant attention!

9. A BRIEF OVERVIEW OF COSTS

A widely discussed point is the issue of analyzing the project's feasibility and project area costs with the consultants and the manager. This explanation does not fit here; for it is worth a dedicated material. But rather a brief overview of these costs for a general understanding.

Regarding the project management area of any company (I won't call it the "Project Office" for not labeling a condition), given the new opportunity, in general 20% of the hours needed for this implementation are considered as due to the effort with management (percentage of hours dedicated to project management). In short, considering the allocations of the most diverse consultants, it is a fact that the allocated team is, in practice, a "fixed cost", although many companies deny this fact! The financial management of this project must consider this account to understand its feasibility.

A project manager on a 40-hour (one week) "symbolic" project was sold for the equivalent of 8 hours (his participation in this one-week project). I don't know any project manager who is available to the client only for these 8 hours, "turning off the taximeter" after this consumption and / or even charging new hours for the need to continue their activities.

The pool of consultants has a fixed cost for the company and the current account must be accounted for, see the sale / allocation in projects during the period of the month to which your salary refers. For example, a project manager who earns $10,000.00 per month, considering 220 hours/month, has an hourly value of $45.45. Considering the hourly sales value of $220.00/hour, practically 1 week sold (45.45 hours) would pay this manager's cost to the office - obviously this hypothetical example is not considering many obligations. It's just for quick understanding! Following this line of reasoning, for 45.45 hours to be equivalent to a percentage of 20%, we are talking about a 227.27 hour project: just over 1 month of project; or little more than an allocation of a consultant during the entire period of the month: A 6 week project (264 hours) considers 20% management (52.8 hours), equivalent to the value of R$ 11,616.00. One week, in hours, is enough to pay this project manager's costs (no other obligations already mentioned).

But, over the course of the schedule, that amount of hours would be spread over the weeks the project takes place and also into the following week, at least for project completion. It is a timeline distribution that will have unfilled gaps if there are no other projects in parallel. You cannot turn off the meter! The manager is the point of contact, the relationship with the customer, the Owner of the analogous Product, who bridges the gap between the team and the customer. He (the GP) will always be available to the customer during the project lifecycle.

Thus, the allocation indicator and the increase in the project portfolio will indicate if the area is "paying itself" or if it is supported by other cost centers in the company.

Managing projects is managing time! And within these times, manage the stakeholders who see your project as unique in their portfolio and you as the professional who is dedicated to this continuous implementation.

10. WHAT IS THE BEST WAY TO MANAGE PROJECTS?

There will always be questions like these, related to a company's PMO (Project Management Office), identification of organizational maturity, among others; calculate the feasibility of a project within the portfolio; and how to support several managers in their activities, which has been very valuable, as there are countless business verticals that are being supported by projects.

Very good that! I see a positive scenario that should extend to all areas, which is a more intimate involvement of professionals with their work. And even though different professionals work in different companies, with different organizational climates and cultures, the doubts are similar.

My biggest request for attention and reflection is in the constant exercise of a change of perspective, leaving the parallel - horizontal vision; for an overhead look - vertical; allowing for spectacular discoveries and contours to achieve the goals. It is important that new and inexperienced professionals have the humility and willingness to manage correctly, open to learning (mutual), to continuous learning! This learning that I refer to is also my learning, with the wealth of variety and challenges.

Managing projects is a continuous learning process, a constant PDCA in the way of managing, a reinvention of posture and behavior in the face of the different scenarios that are presented to us, with the record of lessons learned not only from the activities that took place in a specific project; but also the situations experienced by different managers throughout their careers.

It is the right path to success to have a multidisciplinary team. But what is this? It is nothing more than having a team of professionals from different areas of knowledge that complement each other. The difference in professional profiles and skills allows for a broader discussion of the problem and thus makes the best alternative solutions possible. Best solution alternatives means evaluating from the standpoint of innovation and value delivery.

11. WHAT IS THE BEST TEAM?

IT Governance considers knowing the entire set of intellectual and material needs of the institution, clearly mapping the internal and external customers, the deliverables necessary for its operation and how to guarantee availability and contingency.

The construction of a multidisciplinary team allows specific performance in different business areas (or verticals), but which complement each other in the face of a given project. Diversifying the technical skills, behaviors and experiences of the professionals in a team in a planned way helps to lead the company faster in obtaining results. But for the management of this team to be effective, the leader must carry out strategic plans for the different skills, work each professional in a different way, identifying and developing their talents and thus ensuring the high performance of the team, acting side by side showing support and interest and thus creating motivation.

Promoting team diversity within companies is important and represents a great challenge, as it means managing multiple thinking. Plurality requires knowing how to listen first and foremost. I include here a definition of "horizontal" and "vertical" alignment. In the "horizontal alignment", the leader aims to complement the different intellectual and behavioral skills, enabling greater coverage in the discussion of the problem and, thus, identifying the best alternative solutions. In "vertical alignment", plurality is based on lived experiences. Both require a communication model with the premise of the "listening" exercise.

12. I AM A GOOD LEADER?

The leadership model makes a difference in the face of differences. The model that works best is based on "delegation". When is the right time to point in a direction and let your creativity flow? It is important to note that younger members may view lack of guidance as indifference, while older members may view confidence.

Innovation is "making room for error". The company that innovates, also innovates when choosing leaders. Innovative leadership encourages professionals to take risks. Because if there is no trial and error, there will be no creativity! An innovative environment is a creative environment! In which professional environment are you inserted?

The company usually has the power to decide people's lives. According to Gabriela Prioli, the "boss vs. subordinate" management format favors a complicated relationship and can lead to false responses from the subordinate, motivated by the subsistence factor. False answers that can compromise the performance of this professional and consequently the performance of the company.

Creativity is a value that is increasingly considered a competence; a skill in the hiring decision. Having creative collaborators who think of creative solutions to existing problems is allowing them to make mistakes too! In the current context of world, the fear of unemployment haunts professionals. And afraid of being fired, of losing their job and compromising the livelihood and family support, the employee will naturally give up creativity in favor of security and stability.

The good development of the company is also associated with encouraging a culture of leadership that encourages professionals to explore their own capabilities, thinking about the emotions of employees, allowing them to be creative even if they fail at some point. It is in the company's interest for its employees to be more creative!

If we don't set priorities, someone else will do it for us. Success paradox: If we stop to think, in our society we are punished for the good behavior of knowing how to say "no" and rewarded for the bad behavior of always saying "yes". The first one usually feels weird when it's said! The second is celebrated! This behavior can give the false impression of meeting the requirement of a much-cited professional quality: "availability".

Being "available" does not exactly mean that we must always say "yes" and never say "no", as this only demonstrates ignorance of our own limits, lack of sense of urgency, lack of organization and planning, among others.

In this way, our vision of success is also distorted. The four (predictable) phases below summarize the "paradox of success":

1ST PHASE: When we really have clarity of purpose, we can succeed in initiatives.

2nd PHASE: When we are successful, we gain the reputation of being the person who solves. We are seen as someone who is always available when needed and increasingly presents us with more options and opportunities.

PHASE 3: More options and opportunities, which actually mean more demands on our time and energy, lead to dispersion of efforts. When that happens, we are very overwhelmed.

PHASE 4: We end up moving away from what should be our maximum level of contribution. The effect of success destroys the very clarity that led us to achieve it in the beginning.

Either way, the pursuit of success can be a catalyst for failure. In other words, success can keep us from focusing on the essential things that produce success before anything else..

13. FINAL CONSIDERATIONS

Corporate Governance: There are no isolated implementations! It's amazing and wonderful how everything is related. I usually say that we can take advantage of this and succeed in any objective to be implemented, just knowing these connections, people and processes. The organizations are led, encouraged and monitored by a system that involves, in addition to the board, partners and boards of directors, the relationship with the inspection bodies and interested parties - the stakeholders; aiming to ensure the elimination or reduction of conflicts of interest.

This entire set of regulatory processes, laws, policies and customs define corporate governance, which goes beyond the organization's internal limits and with multiple approaches, maximizing value for shareholders. And it has nothing to do with the Administration, which focuses on the company's internal management.

IT governance is a subset of corporate governance with a focus on information technology, performance, sustainability and risk management; and greatly contributes to the transformation of the company from a functional structure to a designed structure, simpler and more flexible.

Position yourself in front of the new!

A fundamental point in the maturation process of a new idea is the motivation of the people involved! But what permeates the maturation of people's ideas and motivation? Change management!

Going through changes often requires more than a simple decision and achieving success is much more related to the "individual" than to the "collective". According to the ADKAR model, the principle of everything is "awareness" of the need for change!

Awareness of the need for change! It's no use moving forward without overcoming this first step in the face of change to come!

SUCCESS!

www.ingramcontent.com/pod-product-compliance
Lightning Source LLC
Chambersburg PA
CBHW040329220526
45473CB00009B/2625